The
Caledonian Canal

Twinned with
Gota Canal - Sweden
Rideau Canal - Canada

D1340351

AURUM CANAL ROUTE GUIDES

THE
Caledonian Canal

ANTHONY BURTON

FOR BOATERS · WALKERS · CYCLISTS

Aurum Press

To Åke and Pauline Inghammer
with thanks for all their help.

First published 1998 by
Aurum Press Ltd, 25 Bedford Avenue, London WC1B 3AT

Text and photographs copyright © 1998 Anthony Burton

Maps copyright © 1998 GEOprojects (UK) Ltd

Based upon Ordnance Survey 1:25 000 and 1:50 000 maps with the permission of the
Controller of Her Majesty's Stationery Office © Crown copyright (43372U).

The illustration on page 24 is reproduced by courtesy of British Waterways.
The illustrations on page 23 are reproduced by courtesy of the National Portrait Gallery.

A catalogue record for this book is available from the British Library.

ISBN 1 85410 554 X

2 4 6 8 10 9 7 5 3 1
1998 2000 2002 2001 1999

Designed by Robert Updegraff
Printed and bound in Italy by Printers Srl, Trento

Front Cover: *The Caledonian Canal at Corpach.*
Title page: *The eastern end of Loch Lochy.*

Contents

Introduction

The Caledonian Canal

Useful Information

Distance checklist

This list will help in calculating distances between places along the Great Glen and the Caledonian Canal. The distances given relate to the route by water, and walkers and cyclists will find the distances they have to travel, when they vary from the canal route, at the beginning of each chapter. Boaters planning this trip will also need to estimate the time taken to negotiate locks and bridges, and this information is given in the marginal notes.

Approximate distance from previous location

	miles	km
Corpach Sea Lock	0	0
Neptune's Staircase	1	1.5
Gairlochy Top Lock	7½	12.0
Laggan Lock	10	16.0
Aberchalder Bridge	6	9.5
Fort Augustus	4½	7.0
Urquhart Bay	16	22.6
Dochgarroch Lock	10	16.0
Muirtown flight	4	6.5
Clachnaharry Sea Lock	1	1.5

Waterway information

The canal is open on the following days:

Summer	Monday to Sunday
Spring and Autumn	Monday to Saturday
Winter	Monday to Friday

Maximum Recommended Craft Dimensions

	Length	Beam	Headroom	Draught
Caledonian Canal	150'	35'	110'	13'6"
	45.72m	10.67m	33.52m	4.11m
Caledonian Canal	150'	35'	89'8"	13'6"
(with passage under Kessock Bridge on the Moray Firth)	45.72m	10.67m	27.4m	4.11m

The modest lighthouse that stands on the shore of Loch Linnhe marking the western entrance to the Caledonian Canal.

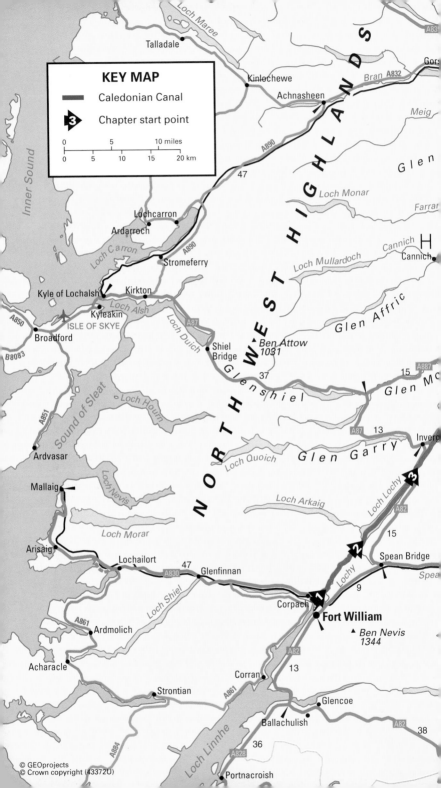

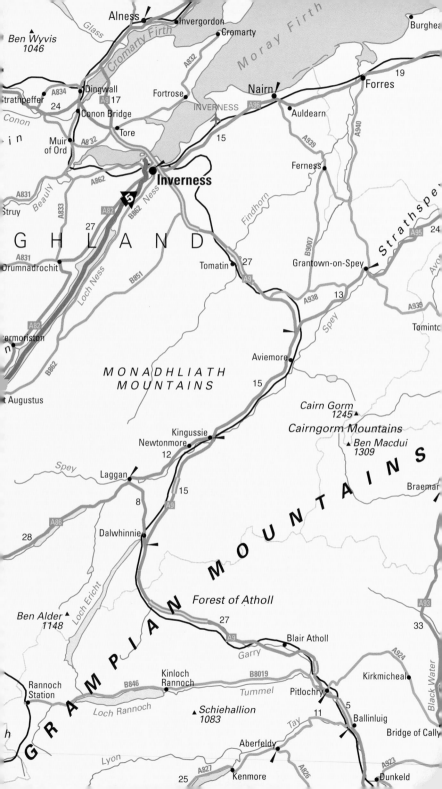

Introduction

The route

This guide is intended for use by boaters, walkers and cyclists. The route for boaters is the simplest, and for the canalised sections the three coincide, with walkers and cyclists making use of the canal towpath. In going around the lochs cyclists follow the Great Glen Cycle Route which is waymarked in the more remote sections by green posts with a bicycle symbol. This route has not yet been fully developed, and it is hoped that at some stage it will be possible to devise alternatives avoiding the busier roads. The way described for walkers also makes use of much of the Cycle Route, but where a footpath is available as an alternative to a road section, this has been used. These paths are not waymarked, and although a full description is given, walkers are advised to take the appropriate maps, especially for the section from Drumnadrochit to the final section of canal leading towards Inverness.

Where the routes diverge, the information for boaters is given first, followed by the cycle route and then the walking route.

Boats on the canal

In order to use the canal, boat owners must obtain a licence, the cost of which depends on the time to be spent on the waterway and length of the craft. Owners wishing to apply for a long-term berth – over 28 days – must comply with the Boat Safety scheme, and they will be subjected to a boat safety check on arrival at sea locks. They must also have appropriate third-party insurance. Hire boats are available on the canal for cruising between Neptune's Staircase and Muirtown Locks. There is a speed limit of 6 miles per hour on canalised sections, but no restrictions on the lochs traversed along the way.

Navigation presents few problems, and where there are hazards these are clearly marked by buoys and warning signs. Moorings are sparse because of shallows at loch sides, and the canal banks slope steeply, making it difficult to come alongside except at wharves and jetties. Mooring places are described in the text and indictated on the maps. Care must always be taken when moving between the natural waterways of the lochs and the artificial canal, and the task is made easier by the system of buoys: the familiar red and green, with red marking the northern edge of the navigation channel and green the southern. More details are given in the main text, where the whole trip is described. Normal rules of navigation apply: oncoming vessels should pass each other port side to port side and overtake another boat by keeping it to starboard. A great variety of vessels use the canal, and the general rule is that power gives way to sail, but this does not apply to large power craft in restricted waters, where there is insufficient room to manoeuvre.

The sea lock entrance at Corpach looking south across Loch Linnhe to the massive bulk of Ben Nevis.

The Caledonian Canal is an inland waterway, but the wide expanses of the lochs can seem like miniature seas and can roughen up quite badly in stormy weather. No one can ever predict weather in the Great Glen more than a day or two ahead. The author has cruised the canal in August and been forced to go and put on a second sweater while taking a turn at the wheel. On other occasions it has seemed, if not tropical, then remarkably hot for northern Scotland. In other words, it is as well to be prepared for anything.

Estimating journey times can be difficult, not least because it is hard to judge how long to allow for passage through locks and swing bridges. These are all under the control of British Waterways officials, who generally prefer to wait for as many boats as possible to fill a lock before starting operations, which may be frustrating for some individuals, but is a sensible policy for conserving water. The times given elsewhere in the book are approximations only. Locks are not operated 24 hours a day, seven days a week throughout the year. They are worked all seven days in summer, closed on Sundays in spring and autumn and closed altogether at winter weekends. Information on opening hours can be obtained from the canal office at Inverness (see Useful Addresses, page 92), together with other details, such as closures for maintenance. Factual information relevant to boaters is given in the margins of the main text.

The handsome lock cottage at Neptune's Staircase, designed by Thomas Telford.

Cyclists

All cyclists who use the canal towpaths must have a British
Waterways licence and comply with British Waterways regulations,
of which the most important on the Caledonian Canal is the
requirement to dismount when passing lock flights. In general,
towpath surfaces are good, give or take the occasional pothole, but
other sections of the Great Glen Cycle Route vary from busy main
roads to very steep rough tracks. The route has not yet been
finalised and it is hoped that some of the road sections will be
avoided in future. The general rule is to follow waymarks as far as
possible.

Walkers

There are no restrictions on towpath walking and none of the
canalised sections present any problems. The routes around the
lochs, however, vary considerably from easy paths at the water's
edge to routes that climb high into the surrounding hills. Those who
intend to follow the whole route should treat this like any other
long-distance walk and come prepared for all eventualities. The
going is rough in parts, so good walking boots are an advantage.
Although the nature of the Great Glen makes it unlikely that anyone
could get seriously lost, life is a lot easier with a good map and
compass. In general, the Ordnance Survey 1:50 000 Landranger
maps are adequate, but the 1:25 000 Pathfinders are probably better
for the walk down Loch Ness. It is a mistake to think of this as being
a level saunter beside a waterway like other canal walks. Parts are
quite demanding, and anyone planning the walk should think about
the amount of uphill slog involved in any one stage. There are long
sections in completely wild country so it is also essential to make
sure that there is enough food and, even more importantly, drink,
available for the day's journey. And because parts of the route take
you into remote country, make sure that someone knows where you
are going and where you expect to finish the day, just in case
anything goes wrong.

It is a very good idea to include a pair of binoculars in the
equipment, for the canal walk is particularly rich in bird life.

The cycle route and footpath overlook Urquhart Bay with the castle on the headland.

Transport in the Great Glen

A casual glance at a map of Scotland suggests that it is cracked in two, broken apart along a line between Loch Linnhe and the Moray Firth, and something very like that did indeed happen, around 350 million years ago. The great tectonic plates beneath the earth's surface shifted, sliding against each other to create a natural fault line. But the landscape we see today owes its shape just as much to what is, in geological terms, a much more recent event, the Ice Age. As the glaciers spread across the land, they bit deep into this great natural fault to create the system of deep lochs and rivers. There are spectacular reminders of those ancient glacial lakes in the 'parallel roads', seen at their finest in Glen Gloy to the south of Loch Lochy, which mark the old shorelines.

This string of lochs, surrounded by high hills and mountains, forms an obvious natural route from coast to coast. No one can be sure just how many prehistoric tracks there are along the Great Glen and how many of them remained in use through later centuries as men and beasts roamed what was once a wild countryside. There are, however, extensive remains of Bronze Age and later Iron Age settlements throughout the region. In more modern times, the area took on the character of the rest of the Highlands. Land was divided between different clans, each owing allegiance to a chief, who administered the region through tacksmen, who held 'tacks' or leases to the land. These groups had little or nothing in common with the town-dwelling people further to the south, or even the rural populations of lowland areas with their neat fields and villages. The clansmen were not just a part of a rural economy based largely on cattle: they were the chief's private army. When, inevitably, the two systems clashed in the Jacobite risings of the late seventeenth and early eighteenth centuries, the whole culture of the region was changed for ever. Much of the pattern of development we see today has to be seen against this background of war and violent change.

The strategic importance of the Great Glen is obvious, and the English set about controlling it through a system of forts. The first crude fortress at the western end was built by General Monk in 1655, then replaced by a more substantial stone structure in the reign of William III and renamed Fort William. It has since been demolished, but the nearby town still carries the name. The equivalent at the eastern end was Fort George, originally established on Castle Hill, Inverness, later moved further west to a promontory overlooking the Moray Firth. In the centre was Fort Augustus, and all three were linked by a military road constructed down the Glen under the direction of General Wade between 1725 and 1726. This was the first true road to be built in the region, and much of it survives either as tracks or part of the modern road system. Walkers follow the old route, still recognisable as a constructed road, down the south shore of Loch Oich.

Forts and military roads helped to bring an end to war in the Highlands, but did little to restore prosperity in peacetime. It was here that the government stepped in to create a wide infrastructure of new roads, bridges and, among the other schemes, promoted the Caledonian Canal. This story is told in the subsequent section.

By the time the canal was open in 1822, steam locomotives were already busy in the collieries of north-east England and, just three years later, the world's first public railway was opened between Stockton and Darlington. The railway age had a profound effect on

The old and the new on the canal: one of the last of the old Clyde Puffers, the cargo ships that served the Highlands and Islands, alongside a modern pleasure cruiser.

The peaceful waters of Loch Lochy with its attractive background of hills.

life on the canals in southern Britain, but the railway builders were understandably reluctant to face the rigours of the Highlands. Inverness, however, was comparatively accessible via the coastal plain, and was finally reached by the Inverness and Nairn Railway in 1855. Progress on the west coast was a good deal slower. The line did not reach Fort William until 1894, and was extended to Mallaig some years later. In the process of building the line, the old fort was demolished, but Fort William found a new identity as a tourist centre.

Both these routes are still open and provide attractive and convenient approaches to either end of the Caledonian Canal. An attempt to extend the West Highland Railway up the Great Glen was not a great success, largely because of the intractable terrain, and the furthest a branch line ever reached was Fort Augustus from a junction at Spean Bridge. It ran out to the pier at Fort Augustus to connect with the popular steamer service on Loch Ness, but it closed in 1933 after less than half a century of use. Walkers encounter the disused line as part of a footpath beside Loch Oich.

Today, modern roads carry the summer holiday traffic and the Great Glen is probably busier than it has ever been throughout its long history.

The canal

The Caledonian Canal was one of the last, as well as one of the greatest, enterprises of the canal age. It was officially opened when a steam yacht crowded with VIPs left Inverness on 23 October 1822, arriving at Fort William on the following day. The story had begun over 20 years earlier. The Scottish Highlands were in an appallingly depressed condition. The scars of the disastrous uprising of 1745 still showed, and the new landlords had no interest in the old communities of the clan system. Thousands were turned off the land and saw emigration as the only answer to their plight. The government was concerned about the loss of manpower and realised that new employment was the key. One promising idea was the development of a thriving fishing industry. New harbours could be built linked by new roads, and there would be immediate work for the builders. One other project was added to the package, a canal down the Great Glen that would enable the fishing boats to cross with ease from one coast to the other without the long and often dangerous

Thomas Telford (left) pictured in old age and William Jessop (right).

passage round the north coast and which would, as a bonus, provide a route for cargo ships and smaller naval vessels. This was to be the Caledonian Canal.

The design of the canal was to be shared between two distinguished engineers, William Jessop and Thomas Telford. Jessop, the senior of the two, was the son of a former shipwright who had become involved in working with the great John Smeaton on the rebuilding of the Eddystone lighthouse. In 1759, Smeaton took on fourteen-year-old William as an apprentice and he remained with Smeaton until 1772, when he set out on his own as a consulting engineer. He was responsible for many important canal schemes, including the Grand Junction, now the main line of the Great Union. It was while working as engineer to the Ellesmere Canal that he employed Thomas Telford as an assistant. Telford came from a very different background, born a shepherd's son in the Scottish Lowlands in 1757. He had worked his way up from being a journeyman stonemason until, in 1786, he was appointed Surveyor for the County of Shropshire. He joined the Ellesmere Canal Company in 1793, and it was there that he made his reputation as the chief instigator of the

A steamer at Banavie showing the original hand-worked locks. The poles were placed in the capstans, and water levels shown by the gauges on the lock gates

daring plan to build the immense 120-foot (36.5-metre) high Pontcysyllte aqueduct over the River Dee near Llangollen, in a feat recorded in verse on a memorial plaque at Clachnaharry. It was Telford who had been asked by the government to report on communications in the Highlands, but the planning of the canal itself was very much a joint enterprise.

The work of building the waterway began with initial surveys in 1802. Jessop and Telford faced particular problems, apart from the scale of the undertaking. Other canal projects of the time could rely on the skill and experience of well organised gangs of navvies and other specialist tradesmen such as carpenters and masons. One of the reasons for the Caledonian's construction, however, was to provide work for the local population, who had no experience whatsoever in this type of work. Whole new villages were built to house the workforce, together with cows in the fields to provide milk to wean the men from the 'pernicious habit' of drinking whisky. No success was ever recorded for this particular scheme. The scale of the whole works was immense. Temporary railways were built, along which stones were carried in trains pulled by horses. Steam engines pumped night and day as the canal was sunk deep below the water level of the adjoining rivers. Dredging machines were also set to work. Robert Southey, the poet, who visited the works with Telford, marvelled at the scene, but was clearly no environmentalist. Of the dredger he wrote: 'Its chimney poured forth volumes of black smoke, which there was no annoyance in beholding.'

Travelling the canal today, one is not always conscious of the scale of the works. The Laggan cutting looks modest in comparison with the surrounding lochs, yet it kept hundreds of men at work, literally for years, and involved diverting the River Lochy down a new artificial channel. The sea lock at Clachnaharry presented even greater difficulties. The seabed of the Beauly Firth slopes very gently, which meant that vessels were unable to get close to shore. So, if ships could not get to the canal, the canal had to go out to reach the ships. A huge embankment of clay was built out to sea, some 400 yards (365 metres) from the highwater mark. Stones were laid on top, and the whole mass was allowed to settle before the channel was cut in the mound and the lock chamber built at the end.

The one great feature that never fails to impress is the interconnected system of locks at Banavie, known as Neptune's Staircase. Southey rhapsodised over the scene, describing it as Britain's 'greatest work of art' and declaring that when set against it, 'the Pyramids

Not the company seal, but an embossed letterhead showing a 'snow', the largest two-masted vessel in regular use in the nineteenth century.

Looking south from the end of the canal at Corpach, with Ben Nevis looming over the village

...erlochy.

would appear insignificant'. Those who came through the locks after the opening may well have been less enthusiastic. The locks were hand operated, using a capstan which required six men to do the work. The lock gates have long since been mechanised, but the old capstans can still be seen.

The canal was never the success its promoters had hoped it would be. At first, few ships of over a hundred tons passed through, mostly taking herring from the Scottish east coast on their way to Ireland. Soon, however, a new type of vessel became a common sight, as steamers joined the old sailing ships, many carrying passengers as well as cargo. The success was short-lived. It soon became clear that cost-cutting during construction had resulted in some very shoddy workmanship, and locks began to crumble just as trade began to grow, with a record 544 ships passing through in 1839. The decision was taken to rebuild the canal and it reopened in 1849, now able to take vessels up to 400 tons, and with a fleet of four steam tugs available for towing sailing ships. A regular passenger steamer service was established, and given royal approval by Queen Victoria who took a trip on *Gondolier* in 1873.

The traffic on the canal has inevitably declined in the twentieth century and now consists almost entirely of fishing boats, yachts and motor boats. Nonetheless, the Caledonian remains a hugely impressive monument to the engineers of the canal age, and it did have one time of glory. Telford had argued that the canal would provide a safe passage in times of war, and during the First World War his point was proved, when literally thousands of naval vessels passed through, mostly carrying mines that were laid to keep German submarines at bay. Telford and Jessop would, however, have surely been surprised to learn that their majestic waterway had found a new lease of life as part of a rapidly growing leisure industry – though they would surely have been delighted to find their great canal thronged with boats, just as they had hoped and planned.

The
Caledonian Canal

Corpach to Gairlochy

via Banavie 8¹/₂ miles (13.5 km)

The approach to the canal by sea from the west has one interesting problem. Loch Linnhe squeezes down to a bottleneck at the Corran Narrows and when a strong tide is running it presents a distinctly uncomfortable passage, with progress against the current very nearly impossible for any but the most powerful craft. All craft will find it a turbulent section, at any state of the tide. Boats proceed up the narrow fjord-like loch to what appears to be a dead end at the paper mill at Corpach, though there is, in fact, an extension of the natural waterway to Loch Eil to the left. The canal entrance is to the east, at the end of the loch, marked by a little lighthouse with a conical roof at the end of a short pier. Vessels should not come alongside when either a red light or a red flag is shown. The entrance lock is generally available at either side of high water and closed

Pump-out, refuse disposal, water point, showers and toilets available at Banavie Top Wharf above the locks.

The Corpach lock-keeper – and other lock-keepers at main flights along the way – can be contacted on VHF Channel 74. The lock-keeper can arrange for customs clearance. A 'licence for time' for using the sea lock and the rest of the man-made canal can be bought at the lock or in advance from the Canal Office in Inverness (see Useful Addresses, page 92).

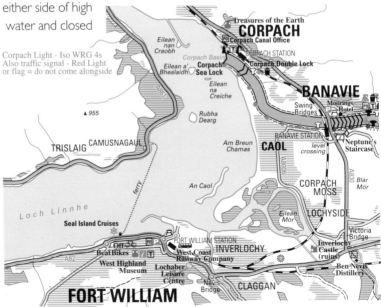

Corpach Light - Iso WRG 4s
Also traffic signal - Red Light
or flag = do not come alongside

▲ 955

The sea lock at Corpach.

All radar scanners must be switched off when approaching or using the locks. Engines should be kept running until all lines have been secured, then switched off to avoid a build up of fumes in the locks.

Average times for locks and bridges:

Corpach Sea Lock
30 minutes

Corpach Top Lock
40 minutes

Neptune's Staircase and swing bridge
90 minutes

Moy Bridge
15 minutes
(bridge is usually left half open,
and small craft go straight through)

Gairlochy Bridge and locks
45 minutes

for 2 hours either side of low tide. Like all the locks on the canal, it is in charge of a lock-keeper, who controls all boat movements. This point is easily reached by walkers from Fort William, on foot, by bus or by train to Corpach Station. The cycle route officially begins a mile up the canal at the A830 road bridge, but Corpach itself is easily reached by taking the B8006 and following the shoreline round at Caol.

The canal gets off to a magnificent start, with the sea lock thrusting out into the water of Loch Linnhe, under the shadow of the brooding, hunched mass of Ben Nevis. Immediately beyond the lock is a wide basin. Yachts and small craft should not moor here but carry straight on up the double lock which gives access to the first long stretch of canal, with the towpath for walkers and cyclists on the right, or south, bank. The artificial nature of the canal is obvious as it runs on a bank above the shoreline before swinging inland above the village of Caol, which offers a variety of shops. Beyond the houses the view opens out to reveal the deep valley of Glen Nevis. Soon two swing

Banavie locks, or Neptune's Staircase, with one of the original capstans in the foreground.

bridges come into view, the first carrying the West Highland Railway extension from Fort William to Mallaig, only completed in 1901. This line, one of the most beautiful and spectacular in Britain, runs regular steam-hauled trips in the summer season, as well as its normal passenger service. If a train is due, there could be a wait of up to 25 minutes before the bridge is opened. Beyond that is the road bridge and immediately afterwards the canal climbs 60 feet (18 metres) through the eight interconnected locks of Neptune's Staircase. Today, boats take an average of an hour and a half to pass up or down the locks, but it all took a great deal longer before the mechanised hydraulic system was introduced for opening and closing the gates. Some of the old capstans can still be seen at the lockside with four slots to take the long poles which were pushed by as many as half a dozen men. Cyclists should note that they are not allowed to ride on the towpath past the locks. A substantial, bow-fronted lock cottage halfway up the flight shows off the architectural skills of Thomas Telford, and is very similar to the canal company offices he designed at Ellesmere on the Llangollen Canal. The small cottage at the top of the staircase also doubles as a post office. The wharf provides good moorings for some distance above the locks.

Although these are the times for passing through the locks, there is a one-way traffic system at Neptune's Staircase, so that if you are unlucky enough to arrive just as a procession is starting in the opposite direction, there will be a 90 minute wait before you can enter the first lock. The other important point to note is that lock closing times are based on the time of the last boat leaving the flight, so to be quite sure of going through the locks on any particular day, one has to be at the first lock 3 hours before the official closing time.

In this section, the canal rides high over the surrounding country, swinging round to follow the natural contours, crossing a number of streams and burns by culvert and aqueduct. The towpath is shaded by broad-leaved trees, and through the gaps there are views out over craggy hills. A break in the trees, with a little cluster of houses down to the right, marks the crossing of the Sheangain Burn on a three-arched stone aqueduct that can be reached by a path down the bank from the towpath. The same path can be followed along to the ruins of Tor Castle, supposedly the fortress of Banquo, famous for his part in Shakespeare's *Macbeth* and the avenue of trees by the canal is known as Banquo's Walk.

As the canal swings in another great curve, the views get even more expansive, with the tiny hamlet of Muirshearlich huddled in

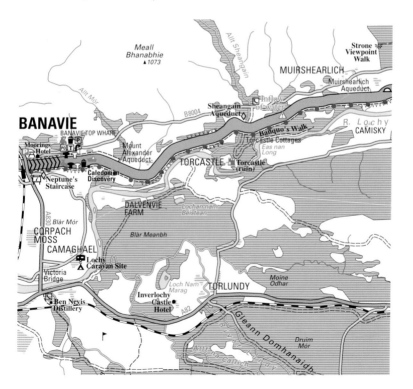

the shade of a forest-covered hill to one side and the dark north face of Nevis, speckled with stark white patches of snow even in summer on the other. White railings stand above a weir taking overspill water down to stone culverts under an impressively tall arch and the River Loy is crossed on a low-arched aqueduct. The broad River Lochy comes into view to the south, wandering languidly over a broad plain, with occasional flurries of activity as it dashes over shallows. The last surviving original bridge is soon reached. Unlike the modern bridges, this is in two halves, and was only designed for use by local farmers. Originally the bridge-keeper would leave his little cottage and open one half of the bridge before rowing across to open the far side – and then repeat the whole process once the vessel had passed. Beyond this bridge, the Moy Burn runs into the canal over a series of weirs designed to trap debris in order to keep the canal from silting up.

The canal near Moy Bridge. The tree-shaded towpath provides an attractive route for walkers and cyclists.

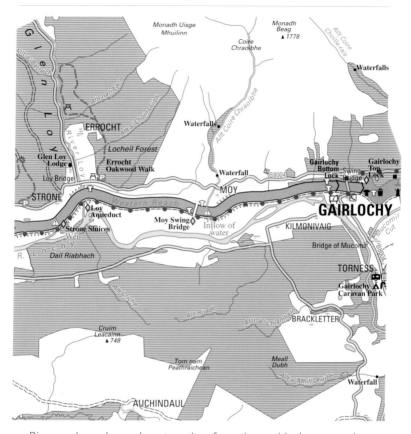

River and canal run close together for a time, with the towpath occupying a narrow strip of land in between, and there is a large overspill weir on the south bank as the canal swings round and the first of two locks that will lift it into Loch Lochy comes into view. Between the two locks is the swing bridge carrying the B8004, and walkers and cyclists leave the towpath at this point and cross over the bridge to follow the road on the north side of the loch. Boats continue on through the second lock into what appears to be a wholly natural waterway, with a small bay on the north bank. In fact Telford had to use the original bed of the loch for his canal, and created a new artificial channel for the river, which can be seen over to the east, where the water pours out of the loch over artificial but none the less romantic falls and is crossed by a handsome Telford bridge. The entrance to the loch itself is marked by a lighthouse of the same design as that at Corpach.

Moy Bridge, like all those on the canal, is a swing bridge and is moved aside to allow ships to pass

2 Gairlochy to South Laggan Locks

via Loch Lochy   10 miles (16 km); walkers and cyclists: 11 miles (17km)

The top lock at Gairlochy, the only one to be built since Telford's day, was constructed by Jackson and Bean's in 1844, following floods in 1834 which saw water running three feet above the lock gates. In 1875, the top gates were raised again and a flood barrier built behind the old lock-keepers' houses.

On leaving the lock, boats must stay in the buoyed channel, keeping well clear of the dam and hydro-electric power station inlet to the south. Once past the Loch Lochy Marina on the north bank, the loch opens out and speed restrictions no longer apply, but this is not quite the entirely natural

Loch Lochy Marina, Gairlochy (tel. 01397 712257): slipway, water point, boat and engine repairs, parts and equipment, cranage, hardstanding.

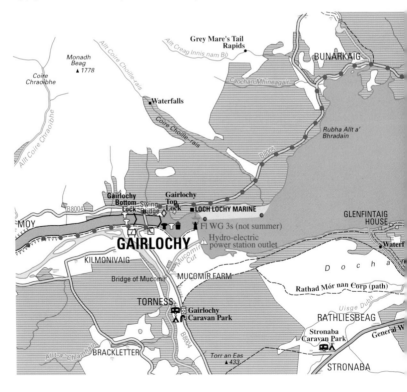

waterway that it seems, for the damming of the River Lochy by the canal builders raised the water level in the loch by 10 to 12 feet (4 metres). It appears, however, to be completely natural and provides Highland scenery at its grandest. Woods stretch down to the water's edge as the hills rise steeply to either side. To the north they climb to craggy summits, the highest topping 3000 feet (914 metres), while those to the south present gentler, more rounded faces. This rugged landscape was a centre for commando training during the Second World War, and there is a commando monument to be seen on the road to Spean Bridge.

At first the loch broadens out to the attractive wide bay of Bunarkaig, before closing in again to a narrow, very deep channel that is typical of the Great Glen. A quiet country road follows the north bank as far as Clunes before heading off into the hills, at which point the busier and noisier A82(T) appears to the south. It follows the line of the old military road, built by General Wade in the eighteenth century to link Inverness and Fort William. There was once a railway squeezed in at the loch side as well, but that closed in 1933.

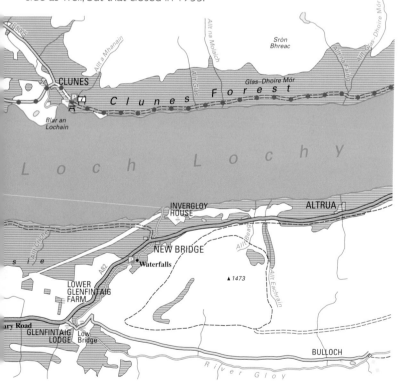

The views remain magnificent throughout the whole length of the loch, and those travelling eastwards should make a point of looking back every now and then to enjoy the superb views over Ben Nevis and the surrounding mountains. A feature of the hills to the north is the so-called system of 'parallel roads' which are not actually roads at all, but long ledges that were once the shores of glacial lakes. As the Ice Age ended and the climate warmed up, the waters shrank down to their present level.

There are few moorings along the loch, but those that are to be found are by suitable

Showers, water point and telephone at Letter-finlay Lodge Hotel.

West Highland Sailing, Laggan (tel. 01809 501234): pump-out, electricity points, boat and engine repairs, cranage, hard standing, water point, refuse disposal. Refuse disposal at Laggan Locks.

Average times for locks and bridges:

Laggan Locks
40 minutes

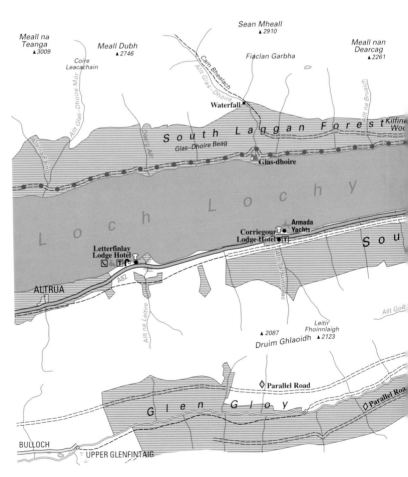

refreshment points. The first, six miles from Gairlochy, is by the Letterfinlay Lodge Hotel, easily seen in a wide clearing where there is a mile-long (1.6 km) break in the woodland; a mile (1.6 km) further on in a second clearing is the Corriegour Lodge. Soon after this, the loch begins to narrow down again to a navigation channel close to the south bank, clearly marked by buoys. The foot of the two Laggan Locks appears virtually at the end of the loch. Here the scenery changes briefly, as the flat floor of the glen is used for farmland. There is a holiday hire base in the little bay immediately before the lock entrance and there are moorings in the bay and at a pontoon jetty.

A pleasure cruiser dwarfed by the grandeur of the Laggan Avenue.

Loch Lochy, the first of the natural waterways along the Caledonian Canal.

Cyclists and walkers

Turn left over the bridge between the two Gairlochy locks, then right at the road junction onto the B8005, which at once begins to twist and turn as it climbs steeply uphill. Although this is designated as a B-road, it is very quiet and carries little traffic. The road soon levels out to run through mainly birch woodland, with occasional glimpses of the loch down below. After a little way, a road to the right leads down to a hire base where those who fancy a trip out on the water can rent a boat for the day. There is a brief conifer-lined section of road before

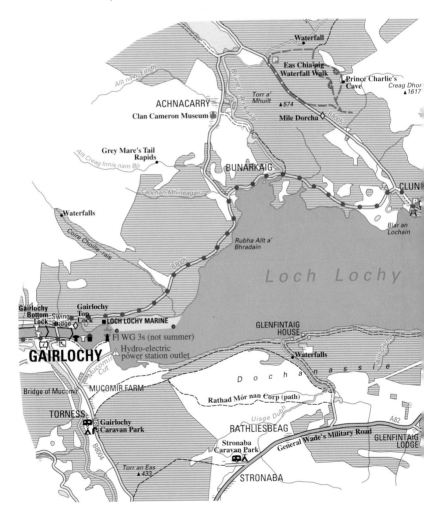

A big European barge, converted for activity holidays, waiting in a lock at Gairlochy.

broad-leaved woodland reappears, with some magnificent copper beech and a variety of mature trees that spread their branches over the road. All along the route little mountain streams can be seen bounding down the hillside and collecting in shallow, clear pools – very useful for cooling the feet on a hot day. As the road slices into the wood, native Scots pine appear with great frequency, standing proud above the other trees.

Eventually the road heads downhill to reach the lochside again, and the view opens up to reveal a craggy hill rising above the bay. A turning on the left leads to the Clan Cameron museum at Achnacarry, seat of the Camerons of Lochiel since the seventeenth century. The road crosses the River Arkaig, which roars down through a series of falls. At Clunes there is a splendid country house opposite a fine old oak with widespread branches. Beyond the house the road crosses a stream. Just after this point, as it begins to swing sharp left, turn right onto the track signposted as the cycle route. At the cattle grid where the way divides, turn right onto the track by the water's edge. This is an attractive path with dramatic crags rising above the lower, pine-clad slopes to the left, and a good view of the softer, grassy hills across the loch. The track climbs through rocks and heather to a fine viewpoint and then promptly descends again to the shore where a shapely hill can be seen standing over the end of the loch up ahead. Once again the route runs through conifer plantations, with bridges over dashing hill streams. On the opposite bank two great gashes in the hillside show the power of these streams to eat deep into the land. There is one more steep climb before houses reappear and the rough track gives way to a surfaced road and a group of holiday chalets. The road runs past an area of small fields grazed by farm animals, crosses a busy burn and heads for the end of the loch. Here the road divides. Cyclists continue straight on along the minor road that climbs up a little hill. Walkers turn right over the cattle grid and take the road past the A-frame chalets to reach the Laggan Locks.

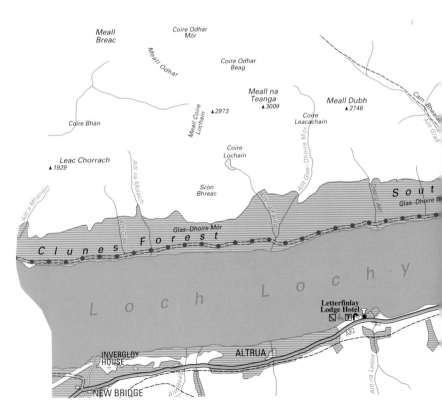

Trees hide the magnitude of the deep cutting at Laggan. This was one of the great engineering works on the canal.

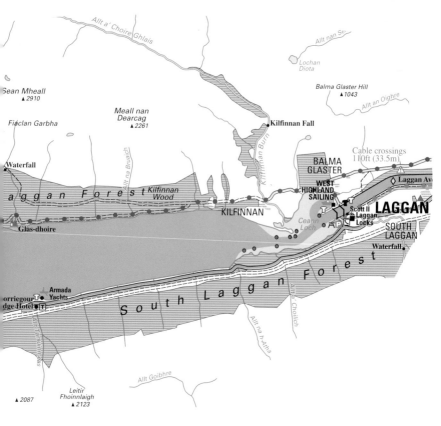

3 Laggan to Fort Augustus

via Invergarry   10½ miles (17 km); cyclists: 12 miles (19km); walkers: 11½ miles (18.5 km)

The next short section of canal proved immensely difficult to construct. Work began in 1814, when hundreds of men made a rough encampment with a ramshackle array of timber houses for overseers and huts for stores. The canal had to be cut deep into the neck of the land separating lochs Lochy and Oich, and the deeper the men dug, using pickaxe and shovel, the harder the work became: as the canal went down, so the spoil rose up as high banks to either side. It was only when the cutting was deep enough to be filled with water that dredgers could be floated in to help. In the end, it was decided to reduce the width at the bottom from 50 feet (15 metres) to 30 feet (9 metres). Even then the work was not completed until 1821.

The Laggan Locks proved equally problematic as they were built on an almost impenetrable foundation of decomposed brushwood. Two artificial banks push out from the shore here to guide boats into the double locks. A number of vessels are moored above the locks, including the old Caledonian Canal ice-breaker, *Scot II*, now serving as a floating pub.

The immense scale of the work at Laggan is not immediately obvious from a boat. The canal is so wide, and the whole scenery of the area is on such a massive scale that the steep banks on either side seem quite modest. It is only when one actually gets a notion of the actual height of these banks, and realises that they are not natural features but the soil and stone dug out of the canal by men using the simplest of tools and brute strength, that one appreciates the immensity of the achievement. The

Head room at Laggan Swing Bridge varies from 9 feet (2.74 metres) to 10 feet 6 inches (3.2 metres). Check with the bridge-keeper before passing, or wait for the bridge to be opened.

Average times for locks and bridges:

Laggan Bridge
15 minutes

Aberchalder Bridge
15 minutes

Cullochy Lock
25 minutes

Kytra Lock
25 minutes

Fort Augustus locks and bridge
60 minutes

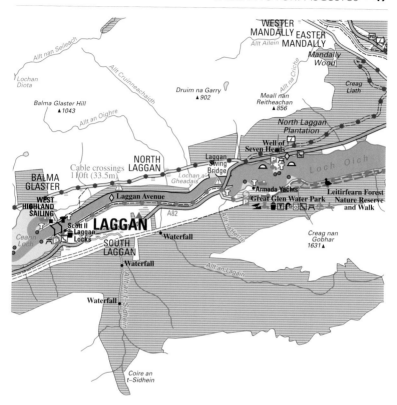

straight tree-lined cutting has been given the rather grand name of Laggan Avenue, and there is indeed a sense of the processional about this part of the route. Boaters should be careful not just to keep to the speed limit, but also to keep wash down to a minimum in this narrow length of canal.

This section ends at the swing bridge, beyond which is the narrowest of all the lochs, Loch Oich, sitting at the top of the Great Glen and forming the summit of the Caledonian Canal. There are moorings to either side. To the south there is also a wide range of facilities at the Great Glen Water Park; to the north a reminder of a particularly grisly historical event. In the early seventeenth century, Keppoch, chief of the Macdonnells, sent his two sons to be educated in France. Whilst they were away, he died and the estate was run by his seven brothers. They had no wish to hand back power, so when the two boys returned they were murdered. Revenge was taken by the family bard, Iain Lorn, who had the seven brothers beheaded and their heads were washed in a spring known today as The Well of the Seven Heads. The spot is marked by a monument; on the lochside beyond are the ruins of Invergarry Castle. It has had a scarcely less turbulent history: the first castle was destroyed in 1654, rebuilt and again destroyed after the Battle of Killiecrankie in 1689. Rebuilt yet again, the castle gave refuge to Prince Charles after Culloden, and as a result the Duke of Cumberland had the castle burned down. There are moorings in the small bay at the mouth of the River Garry, but they need to be approached with caution as there are large expanses of shallow water. In general, Loch Oich presents navigational problems not present in other lochs. Being at the summit, water drains away at either end, and it was necessary to dredge the bottom to create sufficient depth. This was done by

The navigation channel in Loch Oich is marked by red buoys to the north, green buoys to the south. Keep to the channel except when entering the bay at Invergarry. Caution is needed in approaching the moorings and care is also required on the approach to Cullochy Lock. Keep well clear of the wing walls as there are pipes and cables on the bed of the canal.

Pump-out, refuse disposal, water point at Fort Augustus.

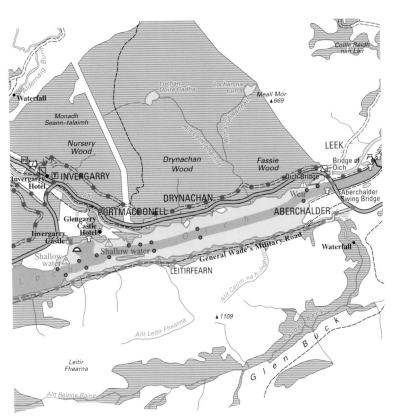

The ruins of Invergarry Castle rising up from the trees on the north shore of Loch Oich.

steam dredgers, one of the first occasions such machines were used in canal construction. The deep-water channel is marked by red and green buoys, and it is important to keep within these limits to avoid grounding. There is a particularly narrow section, where a long, thin, wooded island appears close to the south bank. In general, craft should take particular care anywhere in the loch, not only because of the indicated shallows but also because water levels can vary dramatically throughout the year. The narrow loch has the by now familiar setting of steep wooded banks to either side, with glimpses of rocky outcrops above the tree line. The loch narrows down at the end again, with the River Oich flowing away over a weir on the north bank, and the swing bridge crossing the canal to the south of it. There are moorings to the south. Two bridges can be seen crossing the river: there is a suspension bridge built around 1850 to the patent design of the engineer James Dredge while what appears to be an older, stone-arched bridge is, in fact, concrete, and only dates back to 1932.

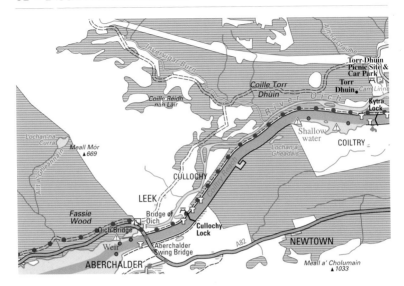

The next section of canal is comparatively uneventful. Cullochy Lock soon appears to mark the start of the descent down to Loch Ness. Beyond this point, the canal runs alongside the river with the hills now set well back. In places, the canal opens out to the south in wide bays which need to be avoided as the water here is very shallow. There is a moment of drama as a high rock face rears up on the approach to Kytra Lock with its charming cottages. Beyond this there are more shallows to be avoided to the south. The glen begins to open out, and the arrival of a golf course indicates that Fort Augustus is near.

The descent by the five connected locks in the centre of the little town is dramatic, and inevitably takes place under the gaze of crowds of tourists – or 'gongoozlers' as they were known to canal families. Once down the locks and through the road bridge there is a short passage out past the stumpy lighthouse for the start of the long haul up Loch Ness. There are, however, ample moorings for those who want to investigate Fort Augustus. This was orginally Kilcumin, the church of Cumin, an Ionian abbot. Then, after the rising of 1715, a fort was built here by the English to control the important pass through the Great Glen. In 1746, the Highlanders captured the fort from the Hanoverian troops and, when the Jacobite Rising had ended, the English garrison returned. By 1867, civil war in Scotland was no longer a threat and the fort was sold to Lord Lovat who handed it over to the Benedictines, who built an abbey on the site. The story is told in a new interpretation centre at the Abbey.

Kytra Lock with its delightful lock cottage, provides a quiet moment before the canal reaches Fort Augustus and the descent to Loch Ness.

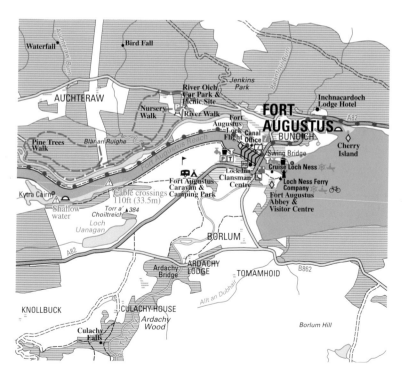

Cyclists

From Laggan the route continues up the minor road, with the canal now out of sight behind its wooded banks of spoil. At the far end of Laggan Avenue the minor road comes down to the busy A82, which is bypassed by a short section of special cycle track leading directly on to the Craig Liath forest road. This climbs steadily and then levels out to run high above the busy main road. Now there are wide views down over Loch Oich and Invergarry castle. The track follows the shoulder of the hill round the mouth of the River Garry before diving down in a precipitous descent to Mandally. At the road, turn right, back towards the loch, and then left on the A82(T) to cross over the river. Turn left onto the A87(T), past the hotel and the tea room, then turn right at the telephone box onto the specially constructed cycle track that zig-zags up the hill through a series of hairpin bends, with views back to the craggy hills above Loch Lochy. This and the other high-level sections are notable not just for their views but also for the wildlife. Large birds of prey are a common slight – buzzards soar high overhead, emitting their odd mewing call, a rather feeble affair for such a majestic bird. Red kites can also be seen, beautiful and fearless birds that seem prepared to fly surprisingly close to the human intruders in their domain. An occasional roe deer might be glimpsed in the wood, and armies of wood ants seem to be regularly on the march down the forest trails. Breaks in the trees provide regular viewpoints down to the narrow loch, where the opposite bank rises precipitously

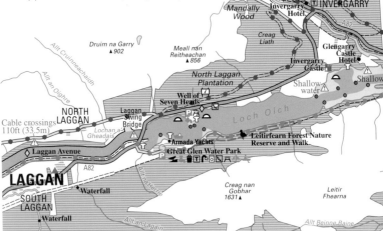

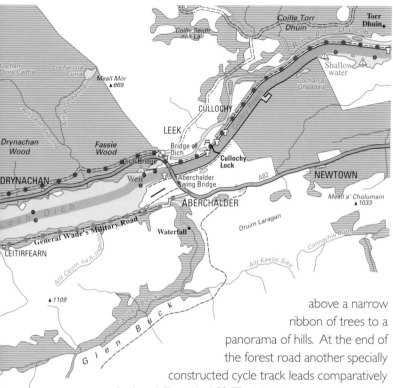

above a narrow ribbon of trees to a panorama of hills. At the end of the forest road another specially constructed cycle track leads comparatively gently downhill to the A82(T). Join the road and continue straight on over the Bridge of Oich to the canal and turn left onto the north bank of the canal to follow the towpath down to Fort Augustus.

Walkers

Walkers can use the route described above for cyclists, but they have the option of taking a more direct – and considerably easier – route, which has the advantage of staying much closer to the water's edge.

From Laggan Locks follow the towpath on the south side of the canal, past *Scot II*. Where this way divides, take the path to the right leading up to the top of the bank. Climbing to the top reveals just what a great mass of material was excavated in making the canal and dumped here at either side. Now the banks are covered in trees and bright with rhododendra, and provide a good viewpoint for seeing the busy passage of boats along the deep cutting. The path eventually reaches the road, close to the Laggan Youth Hostel, a short walk away to the right. To

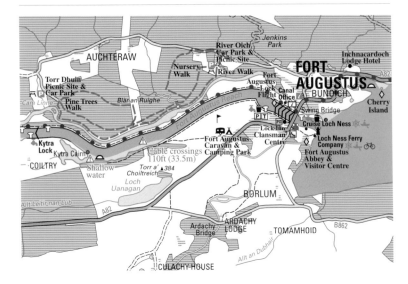

continue along the canal, turn left on the road, cross a little bridge, then turn left again to continue on the footpath. This eventually peters out again at the road at the end of Laggan Avenue. Follow the road round back towards the canal, then turn right before reaching the bridge, on the minor road signposted to the Great Glen Water Park. The road goes through an area of marshland speckled with the bright yellow of irises, passes the edge of the water park and then comes to an end.

Turn right onto the footpath which runs between rhododendron bushes, passing an iron bridge on the right, and very shortly the path joins the trackbed of the former railway line from Spean Bridge to Fort Augustus. The old platform on the right is all that remains of the somewhat inconveniently situated Invergarry Station. The walk runs through the Leitirfearn Forest Nature Reserve, particularly valued for its ash and elm woodland, and rich undergrowth of fern and liverwort. The path divides, offering a choice of routes: continuing on the old railway, or taking a somewhat rougher, narrower path at the water's edge. The two run side by side for a considerable way, with numerous links. Eventually, however, the railway path deteriorates and the walker has to take the lochside route. This arrives at a wooden gate, beyond which the track becomes broader, with evidence of a stone surface. In fact, one old transport route has been exchanged for another, for this is General Wade's eighteenth-century military road. There is a good view across the loch to Invergarry Castle, standing gauntly out from the surrounding trees. The railway was carved out of a narrow ledge, but is now much decayed – an isolated brick chimney is all that has survived of an old line hut.

Nearing the end of the loch, the track climbs back up to the railway, which dives through a cutting with moss-covered rocks to either side. Leave this section by a stile, while the railway track continues on over an iron girder viaduct across the Calder Burn. The path also crosses the burn on a footbridge, after which it continues past the farm to the road. Turn left on the road and cross the canal to turn right along the towpath to continue on the route already described down to Fort Augustus.

The staircase at Fort Augustus crowded with boats. A popular place for onlookers, or 'gongoozlers' as they were known on the canals.

The view down Loch Ness towards Fort Augustus.

4 Fort Augustus to Dochgarroch

via Invermoriston and Urquhart

26 miles (42 km); cyclists: 32 miles (51km); walkers: 28 miles (45.5 km)

The navigation of Loch Ness presents no difficulties. It is a mile wide and 700 feet (over 200 metres) deep in places and there are no records of it having frozen. It can, however, be distinctly rough when the wind funnels down the narrow valley, and if the wind happens to be north-easterly it can be uncomfortably cold even in mid-summer. Nowhere do the seasons and weather do more to alter the mood of a journey down the Great Glen than here. When the sun shines, the water sparkles and the rocky summits positively gleam. When storm clouds gather the water turns sullen, seeming almost oily in its sleek darkness, and the hills become dark, brooding masses. It is on the latter days that it is easy to believe the waters might indeed be home to a monster, a survivor from some distant age. 'Nessie' has, it seems, been around for a very long time. St Adamnan of Iona reported *aquatalis bestia* (a water beast) back in the seventh century, but modern interest was aroused by an alleged sighting in 1932.

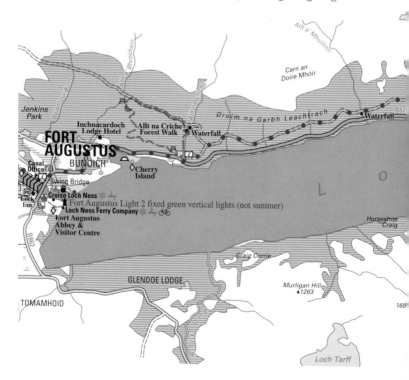

Since then there have been underwater surveys and regular watches, but no-one has yet been able to identify the loch's most famous inhabitant.

Leaving Fort Augustus, the first point of interest is Cherry Island, close to the north shore, with floating moorings close by. This is not, in fact, a natural island, but a 'crannog' built in the Iron Age. Stones were heaped on the bed of the loch to form a platform for timber houses, providing a certain amount of security from attack. In general, there is now deep water almost up to the shoreline, with the exception of shallows off the point of Portclair, some five miles (8 km) up the loch on the north bank. Invermoriston, which appears shortly afterwards, is an attractive spot, but the approach by large craft is prevented by shallows.

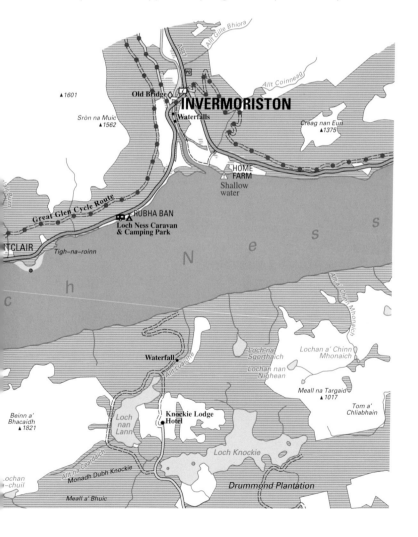

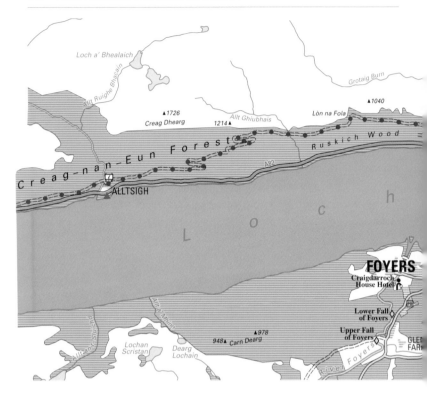

There are moorings to the south in Foyers Bay. This is an interesting spot for industrial archaeologists, for there are extensive remains of the aluminium smelter established there in 1896, the first in Britain to use hydro-electric power. The works, although extensive, were built of local stone to blend as unobtrusively as a major industrial plant can do into its beautiful surroundings. The smelter closed in 1967. A village was built for the workforce and an extensive quay was provided to serve the works. For most visitors, however, the main attractions are the magnificent Foyers waterfalls, the first 30 feet (9 metres), the lower 90 feet (27 metres) high. There are further moorings at Inverfarigaig, where there is a Forestry Commission Visitor Centre and forest walks, including one that climbs high in the hills to lonely Lochan Torr an Tuill.

Over on the north bank a mile before Urquhart Castle is a cairn built as a monument to John Cobb, who was killed on Loch Ness during an attempt on the world water-speed record in 1952.

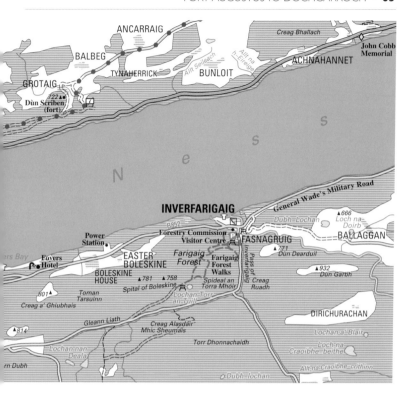

A moody day at Loch Ness, viewed from the cycle track.

Urquhart Castle itself stands proud above Strone Point, a splendidly romantic ruin of towers and turrets in an equally romantic setting. It is not surprising that it is a popular spot for tourists, who in summer queue not just for the castle, but for a place in the car park as well. Boaters should note that Temple Pier opposite the castle is private and should make use of the short-stay marina nearby. Overnight mooring can be arranged, but there is a charge.

There was a vitrified fort here in the Iron Age, but the present castle is a closer relation to the Norman fortress, which consisted of a motte and two baileys. Its strategic importance ensured it a violent history. At the time of Robert the Bruce it changed hands four times in a dozen years. It was repaired during the reign of James IV, but

Slipway, water point, refuse disposal and telephone at Urquhart Bay Harbour, approached down marked channel: red markers to the west, green to the east.

Slipway, water point, refuse disposal, toilets and showers at Clansman Hotel, north shore of Loch Ness.

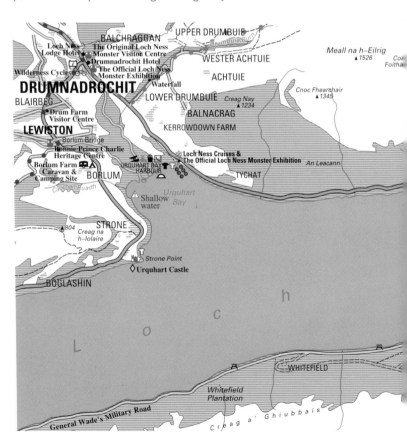

was sacked in 1545 and again in 1644. Its fate was sealed in 1691 when it was blown up to prevent it falling to the Jacobites, and any chance of its being restored as a fortress during the 1715 rising was ended when the weakened walls were flattened in a great storm.

Beyond the castle is Urquhart Bay, with a good deal of shallow water to the west, but a clearly marked deep-water channel to the harbour at the far side of the bay. From here the road leads up to Drumnadrochit, an attractive village built round a green with a number of attractions, including the Official Loch Ness Monster Exhibition and the original Loch Ness Visitor Centre. There are further chances of mooring before the end of the loch is reached, at the Clansman Hotel on the north bank, and at the village of Dores to the south.

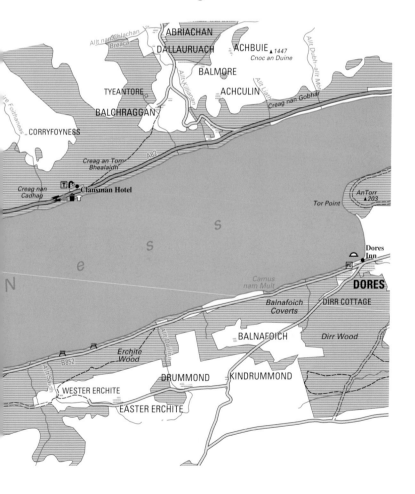

Beyond Dores, the loch narrows down to a buoyed channel that keeps well clear of the south shore. The narrow inlet that gives access to Loch Dochfour is overlooked by Bona lighthouse. When it was originally constructed the lighthouse-keeper must have found sleeping problematic, for the paraffin or carbide lamp was in his bedroom. This was the last manned lighthouse on Britain's inland waterways, but is no longer in use: navigation lights now being

Boats moored up by the lock at Dochgarroch, at the start of the final man-made section of the Caledonian Canal.

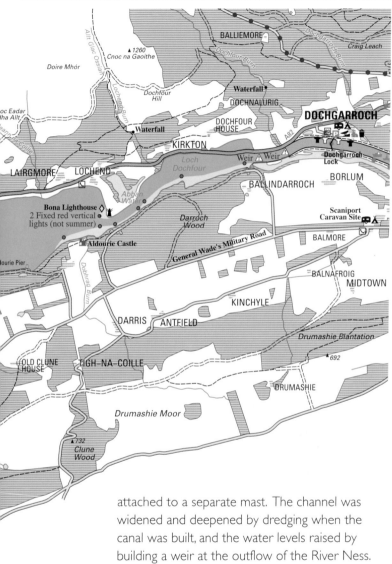

BALLIEMORE

Craig Leach

Dochgarroch Burn

Dochfour Burn

Allt Glac Oisein

▲ 1260
Cnoc na Gaoithe

Doire Mhór

Lochend Burn

Dochfour
Hill

oc Eadar
tha Allt

ian na Coille

Waterfall

DOCHNALURIG

DOCHFOUR
HOUSE

DOCHGARROCH

A82

Waterfall

KIRKTON

Loch
Dochfour

Weir ⚠ Weir

**Dochgarroch
Lock**

LAIRGMORE

LOCHEND

BORLUM

BALLINDARROCH

Abban
Water

Bona Lighthouse ◇
2 Fixed red vertical
lights (not summer)

Darroch
Wood

**Scaniport
Caravan Site**

ourie Pier

Aldourie Castle

General Wade's Military Road

BALMORE

Dobhrag Burn

BALNAFROIG

MIDTOWN

KINCHYLE

DARRIS

ANTFIELD

Drumashie Plantation

OLD CLUNE
HOUSE

TIGH-NA-COILLE

▲ 692

DRUMASHIE

Drumashie Moor

▲ 732
Clune
Wood

**Pump-out, slipway,
water point and
refuse disposal at
Dochgarroch Lock.**

attached to a separate mast. The channel was
widened and deepened by dredging when the
canal was built, and the water levels raised by
building a weir at the outflow of the River Ness.
Loch Dochfour itself scarcely registers as a
separate stretch of water, being little more than a
mile long. A certain amount of care is needed
here, with shallows marked by buoys and weirs to
the south, before the extensive moorings at
Dochgarroch Lock are reached, marking the start
of the final canalised section.

Footpath in the hills above Loch Ness.

Cyclists and walkers

At the end of the towpath, turn left onto the A82(T) to cross the bridge over the River Oich. Poking out of the river is the surviving pier of the railway bridge – the last remnant to be met of the line to Spean Bridge. The cycle route stays with this busy road for a mile and a half, but it is possible to avoid a good part of this, without adding to the length of the journey, by turning left at Bunoich Brae, and following this quiet road round to the right until it rejoins the A82(T) near the Inchnacardoch Lodge Hotel. Turn left and follow the main road past Cherry Island and then leave the road at the Allt na Criche car park. Join the waymarked cycle track that bends round to the right and begins to wind uphill through mixed woodland, dominated by isolated but majestic Douglas fir. Soon pine and other conifers come to dominate the scene, but a lot of woodland has been felled, providing splendid views over the loch to the hills beyond. There are the familiar mountain streams, dashing down little rocky gulleys, one of which drops in a fine, high waterfall. Across the loch is an equally dramatic view of the tall rock face of Horseshoe Craig. The whole route runs at a high level and undulates through attractive scenery of rocks and heather.

The appearance of the shapely, conical hill above Invermoriston marks, it seems, the end of this first section of forest track. The track swings round the shoulder of the inlet leading up to the village, but instead of dropping down towards the houses, continues at a high level. Where the track divides, carry straight on along the route marked by a green post with a white band at the top. Where the path divides again, continue downhill. The route passes massive crags, studded with pine springing from seemingly impossibly small ledges, then begins to head steeply downhill until it reaches the valley floor.

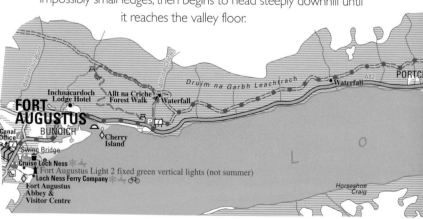

At the bottom, turn right and the track eventually becomes a road passing scattered houses and farms, with fields spreading away to the banks of the River Moriston. After 2 miles (3.2 km) the main road is joined by the road bridge in Invermoriston. Next to the modern bridge is the stone bridge, designed by Thomas Telford, its central pier resting on a little island of rock. Cross over the main bridge and turn left onto the A887, then immediately right past the hotel onto the minor road that climbs steeply through woodland in a series of hairpins – so steeply, in fact, that a warning sign at the top advises cyclists to dismount for the descent. The road eventually levels out, having climbed over 500 feet (150 metres) above Glen Moriston. Cross a small stream, then turn sharp right off the road onto a wide forest track. There is now a respite from climbing as the track heads straight through a typical forestry plantation. Where this comes to an end, there are more superb views over the loch, but then the easy going also comes to a end – an abrupt one.

Turn off the broad track by a marker post and head precipitously downhill, this time on a very rough track, no wider than a footpath.

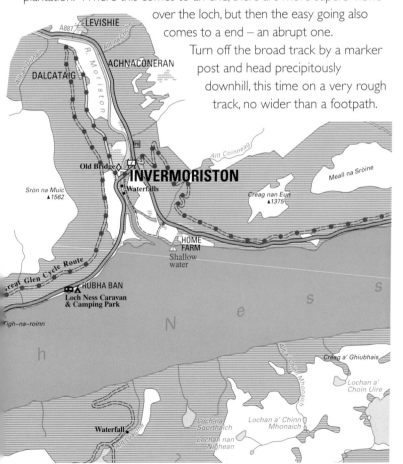

This is part of a series of paths created a century ago by the Invermoriston Estate to help walkers enjoy the superb mountain scenery. Part way along, a little cave has been created under a rock slab, with a stone seat for the weary, or to provide shelter on a bad day. The track levels out again, and now a broad forest track comes into view below, with the main road just in sight below that. This is a very attractive section, along a hillside of rock and heather with wide views and the company of kites and buzzards.

Where the way divides in front of a hillside covered in a dense plantation, take the lower road to the right, which crosses the crashing stream of Allt Saigh, rushing down its narow gorge. Beyond this, there is a turning down to the Loch Ness Youth Hostel at Alltsigh, but the main route continues straight on. Soon the track begins to climb again through dense forest, so that for a time the views are lost. Now there is a long, steady climb until, opposite a radio mast high on the hilltop across the loch, the path turns left and zig-zags through a double hairpin to provide one of the most spectacular viewpoints of the journey so far, looking out not only at the hills that line Loch Ness, but across them to the next range of hills stretching away to the horizon. A final hairpin brings a level track again, but this time hemmed in by trees, and it soon narrows down to a rough, stony path. This ends at a junction with a forest track, but that also quickly narrows again. After a brief downhill section, the track again divides. Take the track to the right and, once through a metal gate, the track runs into woodland,

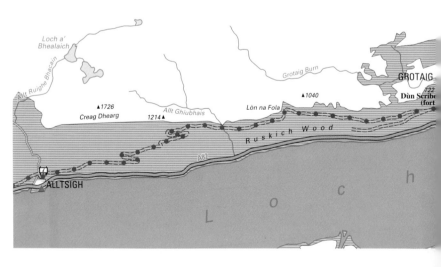

broad-leaved for a change, and dominated by oak. After going through a second metal gate, turn half right by a marker post and take a track that runs down to cross a stream. The woods have now been left behind for farmland. Take the gate on the left and follow the side of the busy stream to the road.

Turn right on the road which runs through fields grazed by cattle and black-faced sheep with immense curving horns. This soon gives way to rougher moorland, studded with gorse and heather, which spreads away northwards to a distant horizon. Eventually the road heads into woodland and steepens as it twists and turns down to Lewiston and the River Coiltie. Turn left onto the A82(T) to cross the river and follow the road round into the popular centre of Drumnadrochit and on to the junction with the A831. There is now a choice of routes. The cycle route remains on roads all the way back to the canal, but walkers who are prepared for some rough moorland walking on what are, at best, ill-defined paths can avoid most of the main road section and save several miles as well. The cycle route will be described first.

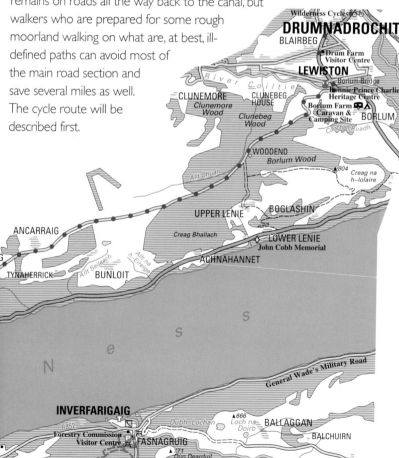

Cyclists

It has to be stressed that this part of the Great Glen Cycle Route is being discussed with the idea of providing a better route, but this is the route in use at the time of writing. Turn left onto the A831, and after 1½ miles (2.4 km) turn right onto the A833 road to Beauly. Inevitably, the road begins to bend and twist as it climbs out of the valley into an increasingly wild landscape of lochans and rocky hillocks. After 5 miles (8 km), turn right onto the minor road signposted to Foxhole, which proceeds through farmland via a series of right-angled bends and at one point runs down a fine avenue of oak. The road climbs again, back to rough moorland, to a T-junction. Turn sharp right and now the views begin to open up over the immense range of hills and mountains to the west and north, and these views just go on getting better and better, wider and wider. At the next road junction, take the road to the right, signposted Drumnadrochit via (rather alarmingly) Loch Ness. This runs down to a forestry plantation on the right. After a mile, a broad, straight track through the trees appears and here walking and cycle route rejoin. The remainder of this section is described below after the notes to walkers.

Walkers

At the T-junction in Drumnadrochit, turn right onto the A82(T) and after half a mile turn left onto the road signposted to Drumbuie. This at once begins to climb very steeply alongside woodland where a stream can be heard, but only occasionally seen, in a deep gully. Where the road divides, keep heading up the hill to the left, with a view down to the loch and Urquhart Castle. The surfaced road eventually ends at a cluster of houses. Keep following the track round to the left to cross the Drumbuie Burn and, once across, turn right onto the rough path over the moor, keeping the thin strip of woodland to the right. This is a wild moorland landscape, very boggy in places with little knolls rising up above the heather. Topping a rise, Loch Glanaidh comes into view, and the road goes past it following the line of the wire fence. The path is very indistinct in places but continues to follow the fence as far as a wooden gate. Go through the gate. There is now no path at all, but a very obvious landmark can be seen ahead and to the right, a whitewashed farmhouse. Head straight for the house.

For a while the going is very rough over peaty ground with dense heather cover,

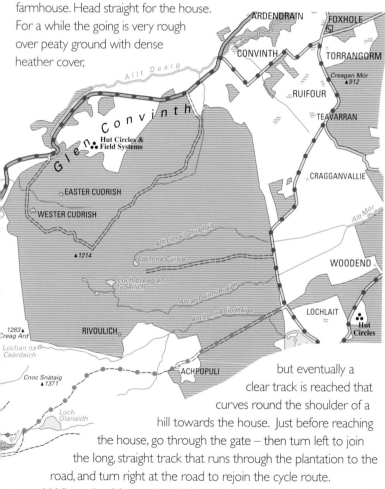

but eventually a clear track is reached that curves round the shoulder of a hill towards the house. Just before reaching the house, go through the gate – then turn left to join the long, straight track that runs through the plantation to the road, and turn right at the road to rejoin the cycle route.

Walkers should note that this is rough ground, with unclear paths and no waymarks. The route is shown in more detail on the OS maps, Pathfinder 192, Loch Ness (North), or Landranger 26, Inverness.

Cyclists and Walkers

The road continues down to the lovely, tree-shaded Loch Laide. Turn sharp left at the road junction, signposted to Inverness. For a while, forest surrounds the road and where tracks appear there are enticing views over the hills. The road finally emerges by the rocks of Lady Cairn, with a string of houses by the roadside, and when those too come to an end there is an immense panorama of hills to the left, dominated by Ben Wyvis rising to a height of 3433 feet (1046 metres), impressive even at a distance of about 20 miles. To the right is a wide expanse of heather moorland, with Loch Ness out of sight over the brow of the hill. As the road begins to descend, the vista of hills slowly disappears and gives way to a no less attractive prospect, closer to hand, of a rough, harsh landscape with rocks pushing up through the heather and a regular spattering of lonely farms. Woodland reappears with fine, mature native pine to either side of the road. This more open forest then gives way to the denser, darker plantations of more recent times, with branches almost meeting across the road which continues on an ever more rapid descent. As the trees clear on the right, there is a view down to the canal and a more distant sight of Inverness.

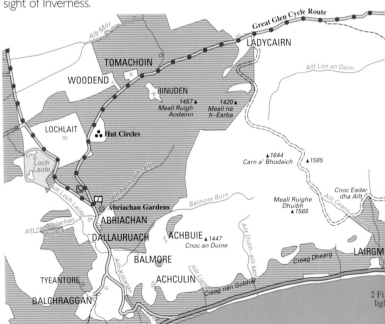

A waymark on the cycle path above Loch Ness.

At this point, the Great Glen Cycle Route turns left onto the main road and takes a direct line into the centre of Inverness. Both cyclists and walkers can, however, now return to the Caledonian Canal and follow it until it meets the sea. To do so, turn right onto the main road and after a mile (1.6 km) turn left by Dochgarroch post office and follow the minor road down to the canal lock.

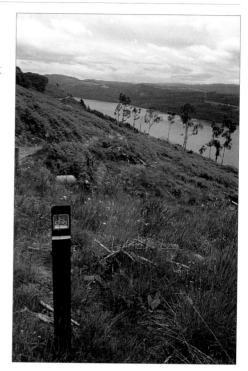

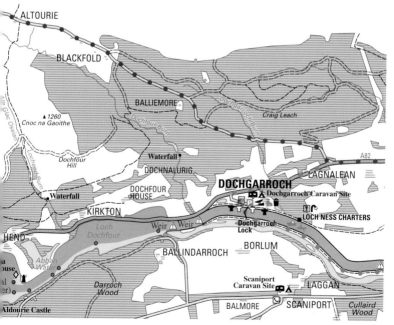

Looking down on Loch Dochfour.

5 Dochgarroch to Clachnaharry

via Inverness   5 miles (8 km)

The remainder of the route consists entirely of artificial canal, running for much of the way alongside the broad River Ness. There are extensive moorings on either side of the canal, but walkers and cyclists should follow the towpath on the south bank. After the drama of the hills, the approach to Inverness seems tranquil, but certainly not dull. This is farming country, with fields full of grazing animals, interspersed with patches of woodland. At first the towpath is tree-lined and provides a very comfortable walk on a grassy track. As the trees open out, the towpath is revealed as a narrow strip of land, between the canal and the generally placid river. The more

As at Corpach, a 'licence for time' and customs clearance can be arranged at Clachnaharry Sea Lock.

Average times for locks and bridges:
Dochgarroch Lock
25 minutes
Tomnahurich Bridge
15 minutes
Muirtown Bridge and Locks
60 minutes
Clachnaharry Works Lock
25 minutes
Clachnaharry Sea Lock
30 minutes

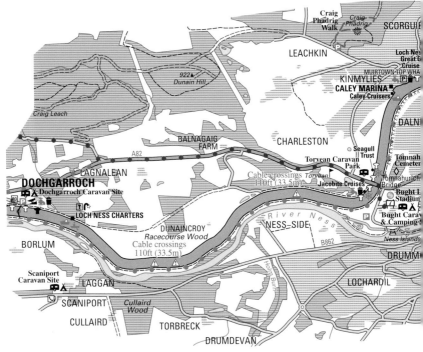

The canal above Dochgarroch. The hills have been left behind and the canal follows the windings of the River Ness.

Times at Dochgarroch Lock may be much longer at weekends, when there is a busy traffic of hire boats leaving and returning to the base.

Tomnahurich and Muirtown Bridges are closed during peak hours for road traffic.

Loch Ness Charter before Dochgarroch Lock (tel. 01463 861303): slipway, water point, refuse disposal, electricity points, bottled gas, diesel, boat and engine repairs, hard standing, parts and equipment and chandlery.

distant view is now of low hills, scarcely more than blips on the horizon. The river becomes temporarily very much more active as it dashes over a long weir, from which water was once diverted to a mill. Shortly afterwards, the river sweeps away to the south, while the canal itself begins a long haul to the north.

The A82(T) crosses the canal on a swing bridge. There is access here to the Bught Sports Centre and Aqua Drome, and, for the more contemplative, there is the surprisingly beautiful Tomnahurich Cemetery, with tombs ranged up the flanks of a shapely hill. Going slightly further afield, a riverside walk arrives at Ness Islands, linked by delightful suspension bridges, built to the cantilever design of James Dredge, very similar to that at Bridge of Oich. The canal now runs through the outskirts of Inverness, still retaining something of its airy, country appeal, with a golf course to the side. The canal swings again to the east to arrive at the four locks of the Muirtown flight, the last of the canal's impressive watery staircases. The road bridge at the foot of the locks provides access to the centre of Inverness down, appropriately enough Telford Street, where a block of fine red sandstone houses was built for the canal contractors to live in. The basin between the bottom lock and the bridge is shallow to the north, but there are extensive moorings in Muirtown Basin just past the bridge, and it is well worth making time to explore the town though not much of old Inverness has survived. Macbeth is said to have had a castle here, but the present castle only dates back to the nineteenth century, as does the cathedral. Most of the buildings are in the Victorian version of Scottish vernacular. What gives Inverness its special character is the broad river, which was carrying

Caley Marina, Inverness (tel: 01463 236539): slipway, water point, refuse disposal, pump-out, electricity points, bottled gas, diesel, boat and engine repairs, cranage, hard-standing, parts and equipment and chandlery.

Seaport Marina, Muirtown (tel: 01463 233140): water point, refuse disposal, pump-out, electricity points and diesel.

Muirtown Locks are the limit of navigation for hire boats.

A plaque at Clachnaharry acknowledging Telford's work on the Goth Canal in Sweden.

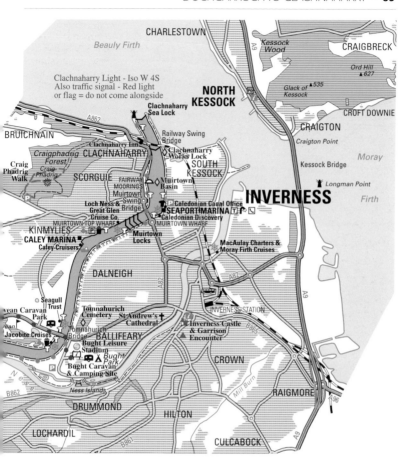

Clachnaharry railway bridge will be closed for up to 25 minutes when a train is due.

The sea lock is normally available for use 4 hours either side of high water. At low water and spring tides the sea lock is closed for 2 hours either side of low water.

commercial traffic long before the canal was even considered and continued in use after it was completed. There are quays along both banks, and the harbour was rebuilt in 1846.

The canal towpath changes sides at Muirtown Basin which eventually narrows down to reach Clachnaharry Works Lock. This was one of the busiest places on the canal during construction, and it remained an important centre for maintenance, so that a small village developed around the various workshops. A fresh problem appeared when the Highland Railway built their line north from Inverness

The neat little signal box beside the swing bridge which carries the Highland Railway line across the canal at Clachnaharry.

in the 1860s. They had to build a swing bridge across the canal and this was replaced by the present bridge in 1909. Movement on both railway and canal is controlled from the little signal box. The buildings at Clachnaharry also serve as a monument to the engineering genius of Thomas Telford. As well as Southey's verses, recorded on a plaque, there is also a permanent record of another great ship canal for which he was a consultant engineer, at Gotha in Sweden.

Beyond Clachnaharry, the canal is thrust out into deep water in the channel built into an artificial embankment, ending at the sea lock. As at the western end of the canal, there is a small lighthouse and a system of traffic lights and flags controlling the approach of vessels from the Moray Firth. Boats leaving the canal for the sea at this point will find an attractive coastline and a welcoming population of seals and dolphins. Those who have travelled the route on foot or by cycle can only turn back inland, their journey over.

The end of the canal as it reaches the sea at the Beauly Firth.

Useful Information

Transport

Information on transport can be obtained from all the Tourist Information Centres (see page 92).

Inverness is a main line station; there are services from Glasgow and Edinburgh, with connections to all other UK cities. Fort William is on the former West Highland line from Glasgow to Mallaig.

Inverness Airport offers fast connections to Glasgow, London and other regional airports throughout the UK.

Rail enquiries:
National Rail Enquiries. Tel: 0345 484950.

Coach enquiries:
Scottish City Link. Tel: 0990 505050.
National Express Coaches City Link. Tel: 0141 443 9191.

There is a bus service between Fort William and Inverness: Highland Country Buses. Tel: 01397 702373.

Boat Trips and Boat Hire

Hotel boats/activity boats	*Departure point*
Caledonian Discovery. Tel: 01397 772167.	The Slipway, Corpach
Loch Ness & Great Glen Cruise Company. Tel: 01463 711913.	Muirtown Top Lock, Canal Road, Inverness
Highland Steamboat Holidays Ltd. Tel: 01546 510 232.	Corpach/Inverness

Holiday boat hire	*Departure point*
Loch Ness Charters. Tel: 01463 861303.	Dochgarroch Locks
Caley Cruisers. Tel: 01463 236328.	Canal Road, Inverness

West Highland Sailing.
Tel: 01809 501234.
Armada Yachts.
Tel: 01397 700008.

Laggan Locks

Great Glen Waterpath Jetty.

Public boat trips

Cruise Loch Ness.
Tel: 01320 366221/366277.
Loch Ness Ferry Company.
Tel: 01320 366233.
Seal Island Cruisers.
Tel: 01397 703919.
Official Loch Ness Monster
Exhibition Centre.
Tel: 01456 450218.
Loch Ness Cruises.
Tel: 01456 450395.
MacAulay Charters.
Tel: 01463 225398.
Moray Firth Cruises.
Tel: 01463 717900.
Jacobite Cruises.
Tel: 01463 233999.

Departure point

Swing Bridge, Fort Augustus

Fort Augustus Abbey

Town Pier, Fort William

Urquhart Harbour,
Drumnadrochit

Urquhart Harbour,
Drumnadrochit
Shore Street Quay, River Ness,
Inverness
Shore Street Quay, River Ness

Tomnahurich Bridge,
Glenurquhart Road

Boat trips for the disabled

Seagull Trust.
Tel: 01463 703456.

Departure point

Tomnahurich Bridge, Inverness

Day boat hire

Loch Lochy Marine.
Tel: 01397 712257.
Cruise Loch Ness.
Tel: 01320 366221/366277.
Loch Ness Ferry Company.
Tel: 01320 366233.
Loch Ness Charters.
Tel: 01463 861303.

Departure point

Gairlochy Top Lock

Swing Bridge, Fort Augustus

Fort Augustus Abbey

Dochgarroch Locks

Accommodation

Information can be obtained from the Tourist Information Centres (see below) on all types of bed and breakfast accommodation. They can also advise on bunkhouse, hostel and camping facilities.

Youth Hostels are at:

Glen Nevis	GR 127718	(01397 702336)	
Loch Lochy	GR 293972	(01809 501239)	South Laggan
Loch Ness	GR 457191	(01320 351274)	Glen Moriston
Inverness	GR 667449	(01463 231771)	1 Old Edinburgh Road

Further information can be obtained from the Scottish Youth Hostels Association at 7 Glebe Crescent, Stirling FK8 2JA. Tel: 01786 451181.

Tourist Information Centres

Scottish Tourist Board, 23 Ravelston Terrace, Edinburgh EH4 3EU.
Tel: 0131 332 2433.
Highlands of Scotland Tourist Office, Grampian Road, Aviemore.
Tel: 01478 810363.
Fort Augustus (seasonal), The Car Park, Fort Augustus.
Tel: 01320 366367.
Fort William, Cameron Centre, Cameron Square, Fort William
PH33 6AJ. Tel: 01397 703781.
Inverness, Castle Wynd, Inverness IV2 3BJ. Tel: 01463 234353.
Spean Bridge (seasonal) Spean Bridge PH34 6DX. Tel: 01397 712576.

Useful addresses

British Trust for Ornithology, Beech Grove, Tring, Herts HP12 5NR.
British Waterways, Caledonian Canal Office, Seaport Marina,
Muirtown Wharf, Inverness IV3 5LS. Tel: 01463 233140.
British Waterways Scotland, Canal House, Applecross Street, Glasgow
G4 9SP. Tel: 0141 332 6936.
Cyclists Touring Club, 69 Meadrow, Godalming, Surrey SU7 3HS.
Tel: 01483 417217.

Forest Enterprise, 231 Corstophine Road, Edinburgh EH3 5RA.
  Tel: 0131 244 3101.

Forest Enterprise, Strathoich, Fort Augustus PH32 4BT. Tel: 01320
  366322.

Forest Enterprise, Torlundy by Fort William PH33 6SW. Tel: 01397
  702184.

Highland Council, Glenurquhart Road, Inverness IV3 5NX. Tel: 01463
  702604.

Historic Scotland, Longmore House, Salisbury Place, Edinburgh
  EH9 1SH.

National Trust for Scotland, 5 Charlotte Square, Edinburgh EH2 4DU.
  Tel: 0131 226 3922.

Ordnance Survey, Romsey Road, Maybush, Southampton SO16 4GU.

Ramblers Association (Scotland), 23 Crusader House, Haig Business
  Park, Markinch, Fife KY7 6AQ. Tel: 01592 611177.

Royal Society for the Protection of Birds and Scottish Wildlife Trust,
  Cramond House, Kirk Crammond, Cramond Glebe Road,
  Edinburgh EH4 6NS.

Scottish National Heritage, Battleby, Redgorton, Perth PH1 3EW.
  Tel: 01738 627921.

Scottish Youth Hostels Association, 7 Glebe Crescent, Stirling FK8 2JA.
  Tel: 01786 451181.

Ordnance Survey maps covering the route

Landranger (1:50 000)

26 Inverness
34 Fort Augustus
41 Ben Nevis

Pathfinders (1:25 000)

177 Inverness and Culloden Moor
192 Loch Ness North

193 Daviot
208 Mid Loch Ness
222 Glen Moriston
223 Invermoriston (Loch Ness)
238 Fort Augustus and Loch Garry
251 Laggan (Highland)
264 Glen Loy
265 Spean Bridge and Glen Loy
277 Ben Nevis and Fort William

Bibliography

Cameron, A. D., *The Caledonian Canal*, Canongate Academic, 1994.

Feachem, Richard, *Guide to Prehistoric Scotland*, Batsford, 1977.

Hadfield, Charles, *The Canal Age*, 1968.

Hadfield, Charles, and Skempton, A. W., *William Jessop, Engineer*, 1979.

Maclean, Fitzroy, *Scotland*, Thames & Hudson, 1993.

Millman, R. N., *The Making of the Scottish Landscape*, Batsford, 1975.

Rolt, L. T. C., *Thomas Telford: a Biography*, 1958.

Taylor, William, *The Military Roads of Scotland*, David & Charles, 1976.

Thom, Valerie, *Birds in Scotland*, Poyser, 1986.

Thomas, John, *The West Highland Railway*, David & Charles, 1984.

Weir, Tom, *The Scottish Lochs*, Constable, 1980.

Whitton, J. B., *Geology and Scenery in Scotland*, Chapman & Hall, 1992.

Witchell, Nicholas, *The Loch Ness Story*, Corgi, 1989.

Places to visit on or near the route

Glen Nevis Visitor Centre
Old Inverlochy Castle
Nevis Range Cable Cars
Ben Nevis Distillery Visitor
    Centre, Lochy Bridge

Fort William
West Highland Museum,
    Cameron Square
The Jacobite Steam Train
Corpach: Treasures of the Earth,
    Mallaig Road
Achnacarry: Clan Cameron
    Museum
Loch Oich: Well of the Seven
    Heads
Fort Augustus
Abbey and Fort
    John Cobb Memorial

Urquhart Castle (Historic
    Scotland)

Drumnadrochit
Official Loch Ness Monster
    Exhibition
Loch Ness Lodge Visitor Centre
Original Loch Ness Visitor Centre
Official Bonnie Prince Charles
    Heritage Centre

Dochfour Gardens
Inverness
Inverness Museum and Art
Gallery, Castle Wynd
Castle Garrison
St Andrew's Cathedral, Ness Walk
North Kessock Dolphin and Seal
Centre

Town plans

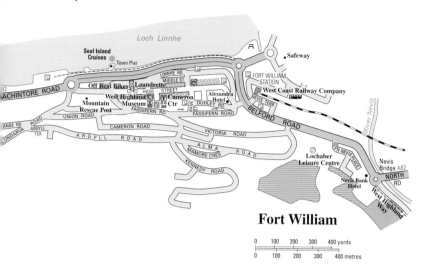

Fort William

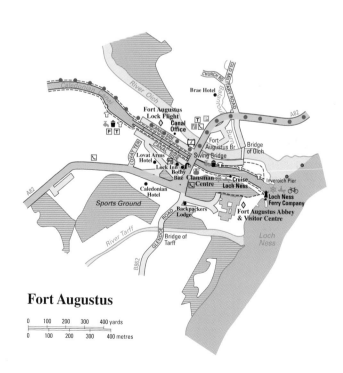

Fort Augustus

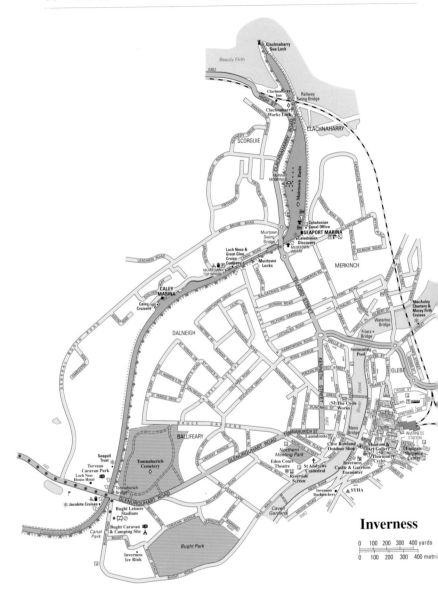

Inverness

| 0 | 100 | 200 | 300 | 400 yards |
| 0 | 100 | 200 | 300 | 400 metre |